The Write Tool to Teach Algebra

Virginia Gray

KEY CURRICULUM PRESS
Innovators in Mathematics Education

Editors: Joan Countryman and Dan Bennett
Cover art and graphics: Susan Malikowski, Autographix
Calligraphy: Virginia Gray

Published by Key Curriculum Press, 1150 65th Street, Emeryville, California 94608.
editorial@keypress.com
http://www.keypress.com

Printed in the United States of America   11 10 9 8 7   04 03 02 01   ISBN 1-55953-064-2

Dedicated to

Vera Mahanay

who taught me that writing could teach

About the Author

As a teacher certified in both Math and English, Virginia Gray has a keen interest in using writing to develop depth in math students. She earned a Bachelor of Arts in English and American Literature at the University of California in San Diego. Her Masters of Arts Degree from Southern Oregon State College is in the interdisciplinary study of Mathematics and English Language Arts.

Virginia lives with her daughter Brenda and her son Kenneth in Ashland, a small city which is tucked between the Siskiyou and Cascade Mountains in southern Oregon. She has taught at both the college and high school levels, and is convinced that mathematical understanding should be developed with a variety of tools in a humanistic manner.

Introduction

Entering mathematics education amidst cries of "Write to Learn," I thought, "What do you mean?" I had personally used writing when trying to sort out complexities in advanced math classes, but I thought that was private. I believed in using writing to teach math, but the question was, "How?"

I tried and flailed. My first effort was a disaster. I asked basic math students to explain why it was acceptable to "carry" in an addition problem. They had no idea! They had no way to figure it out! I had insurrection on my hands! My degree and teaching credential in English had given me confidence to try writing but I found that math class was quite different. I had no model. Math class was steeped in tradition. I had no idea of how using writing to teach math would look. The search began.

Tackling the task with fervor, I collected ideas by asking teachers, reading, reading, reading, and thinking. There was little information. General articles on Writing to Learn usually included the ubiquitous "Have them write their own word problems." Students in my class wrote boring word problems which did not include enough information to be solvable. Writing was no fun that way and did not seem to elicit much thinking or learning. Writing math journals was a common suggestion and a good technique. Still, I felt alone in the task as I could find so little specific information on *how* to use writing to teach math.

This book is a series of lesson plans tied to the algebra curriculum. The order is generally consistent with a traditional course. I chose a modeling format because a specific model for how to use writing would have helped me. I offer this book as an example of how one teacher attempts this task. The ideas are an amalgamation of ideas collected through trial, error and discussion with other math teachers. I offer them humbly, sharing my beginning efforts. I hope other teachers try new ideas and share what they find. Maybe together we can figure this out.

Table of Contents

Chapter	Page

1. I Hate Math

Math is often a stressful experience in our culture, and I give students a chance to put their cards on the table and to open up the subject of feelings in math. I start the year with this activity which offers students a chance to complain and air concerns. You may want to skip this. I like it.

Writing Activity
(Teacher Aid 1.1)

I want you to write about three things:

1. What do you like for a teacher to do in math class? How can a math teacher be helpful?

2. What do you hate for a math teacher to do? Is there something I need to know about teaching math?

3. One time in math class. . . . Complete the sentence, telling me something that happened and how you **felt** about it.

Other Details

I always add a few details to this explanation. I assert that I must test them in some way and that there will be some kind of homework, but I am open to suggestion here. If it makes them feel better to say, "Don't have homework; don't have tests," then fine. But I tell them to add something to that thought. I need help from them, and this is their chance to have a say. They may say anything they want; they're just not allowed to use words that will embarrass me.

Grading

I don't grade this one. I read it and file it.

Math Write

Do this when teaching

Don't do this

One time in math class . . .

2. What is Multiplication?

This lesson introduces the use of writing in the math class. Beginning with a pre-writing activity, the lesson takes the student through a step-by-step process. Students usually have a high success rate with this task. They're reluctant to plunge in, but they all arrive successfully on the other shore.

Opening

Copy the following addition problems on your paper.
Solve each of them.

10	37	$6 + 6 + 6 + 6 + 6 + 6 =$
10	37	
10	37	
<u>10</u>	37	
	37	
	37	
	37	
	37	
	37	
	<u>37</u>	

Discussion

What are the answers? (40, 370, 36)

How did you do it? (multiplied)

But I said, "Add," not, "Multiply."

How do you know when you can multiply?

What is multiplication for?

Pre-Writing

Fill in the chart (*Teacher Aid 2.1*) on the overhead, using student words.

Writing/Pre-writing

Tell me what multiplication is for. Now, I know that writing in math class seems unusual, but it's important to be able to explain ideas and to practice communication by writing.

Don't worry about punctuation or spelling. Just put your thinking on paper. It should take about a page.

Sharing

OK, What is multiplication?
Will some of you share what you wrote?
Remember: I give extra credit to anyone who volunteers.

Note: I remind students how hard it is to share writing, that it is difficult to put math concepts into words and that I expect them to listen attentively.

I comment on the writing, but I don't worry about syntactical errors.

Students share about seven of the explanations if there are that many volunteers.

If there are no volunteers, collect the writing and share it yourself by reading aloud. I have never had that problem. Extra credit seems to be enough enticement.

Back to the Chart
(Return to *Teacher Aid 2.1*)

Shall we add or delete anything?
How can we clarify what we wrote?

Roundup

The teacher needs to highlight the fact that multiplication is repeated addition and can be used instead of adding the same number over and over.

Assignment

On the back of your paper, use **exactly** 25 words to tell me what multiplication is. "A" is a word; "the" is a word. I'm requiring exactly 25 words because then you are forced to choose your words carefully.

Grading

1. Count the words. If there are not 25 exactly, ask for another try.

2. Read the paper. If the idea of repeated addition is not there, reject it and ask for another try.

3. Everyone else gets an A or a B.

Follow-Up

After all students have a correct 25 word explanation, they copy them carefully in ink on a small card. I post the cards so that students can see how others tackled the task.

On the next test, which I give **soon** after this activity, I include the following question:

Using 25 words or less, what is multiplication?

This is easier than exactly 25 words and the students should be well-prepared to answer.

Multiplication

What is multiplication for? | If it's just addition, why do we teach it?

When can I use it? | When can I not use it?

 Teacher Aid 2.1

3. Math Symbols

This activity helps students recognize the importance of symbolism in mathematics and helps dispel the mystery of how and why the symbols developed.

Opening

One of the difficulties for many students in math class is the symbolism. It's confusing to be faced with a page of strange symbols that are so hard to understand that you have no hope with the mathematics behind them.

Mathematicians **think** in symbols. When you work with symbols, they become part of your math language and you don't trip over them.

In fact, you have already learned to think in symbols. For example, solve the following problem as quickly as you can:
(Show *Teacher Aid 3.1* and provide some wait time.)

How did you do the problem? What went through your mind when you solved it?

After discussing the thinking process involved, I emphasize that most of us translate this problem into symbols **before** solving it. We do indeed think in symbols.

I then show *Teacher Aid 3.2*, which is a listing of math symbols. Some of these will be familiar to students. We discuss meanings.

I ask students for their ideas on how some of the symbols may have developed.

I tell them that in the sixteenth century Robert Recorde developed the equal sign for his book *Whetstone of Whitte*. He said he chose two parallel lines of equal length because no two things could be more alike.

Writing Activity

1. Divide into groups of three.

2. Invent a math symbol. Tell how and when to use it.

3. Define the symbol carefully; we should be able to use it correctly after having read your explanation and definition.

4. Give five examples of how to use it.

5. Devise a set of four problems to test whether other students can use the new symbol correctly.

After the students define their symbols, I collect the work and share it with another group. That group reads what the first group wrote, sends it back for any necessary rewriting and then explains it to another group. This is a natural editing process that should help to clean up student language and definitions.

Fifty-four minus twenty-nine is equal to what number?

Symbols

0 1 2 3 4 5 6 7 8 9

$+$ $-$ $\times$ $\div$ $\sqrt{}$ $=$

$\leq$ $<$ $\neq$ $>$ $\geq$

...	goes on forever
$\therefore$	therefore
$\pm$	plus or minus
$\sqrt{}$	square root
$\triangle$	triangle
$\approx$	approximately equal to
$\exists$	there exists
$\forall$	for all

Teacher Aid 3.2

4. Adding Integers

Teaching addition of integers visually lets the students develop individual schemes to solve the problems. The final writing activity gives each student a chance to verbalize his/her scheme.

Before the Students Arrive

I bring red helium balloons and blue regular balloons as a visual analogy. The helium balloons (which rise) will represent positive integers; the blue ones (which sink) will represent negative integers. Actually bringing balloons into class is a colorful, interesting touch, but using red and blue chips works well. I have packets of red and blue poker chips available for student use.

Opening

I have two kinds of balloons—red ones full of helium and blue balloons that are the regular kind. Reds go up; blues sink. If a red and a blue balloon were inflated just right, and we attached them, what might happen?

Somebody will think of the idea of perfect equilibrium—that we could place them anywhere and they would stay put. Meanwhile, tie the two balloons together.

If you have the balloons in the room, you can tie the two balloons together and use paper clips or staples to weight them just right for equilibrium.

So, red and blue balloons cancel each other out. If I had five reds and two blues, they would go up just as fast as _____ red ones?

We then do integer addition disguised as balloon problems. We do jillions of them. Students use the packets of red and blue poker chips and quickly begin to develop a scheme to "cancel" three blues with three reds and so forth.

> ## Zero
>
> How can I represent zero using **at least** five chips? Make
> something that would have zero motion using your chips.

A quick scan of the room will show that they've all got it. Share a few, but don't bore
them to death. It is in this activity that we solidify the concept of cancelling—that like
amounts of each color will cancel and have no effect.

Practice
(Teacher Aid 4.1)

This is a teacher aid presenting balloon problems to be done without the poker chips,
yet providing a visual crutch. Students do it very quickly. It's easy for them but
solidifies the concept.

Writing
(Teacher Aid 4.2)

This activity requires that the students write the rules for adding integers. The rules,
disguised as rules for balloon problems, translate easily later. Students see the transition
as obvious.

Editing

I have students work in groups of three. One student reads his/her writing while the
other students try to use the given rules to add integers. They help each other to clarify.

Grading

I read the papers and work with students to develop rule statements that are accurate.
Each student must work until the statement is correct conceptually. I record a check for
everyone upon completion.

Follow-Up

The students copy the corrected rules onto colored paper in ink. Students post these
rules on the bulletin board or place them prominently in their notebooks.

Balloon Problems

Use ↑ and ↓ to indicate rising or falling. Show your work by making dots over a line for ↑ balloons and under that line for ↓ balloons ⁓.

Example —
     ↓4 plus ↑2
     would be ∴⁓ = ↓2

Your turn:
Yes! You must draw the picture.

↑5 plus ↓6 = ⁓ =
↑2 plus ↑3 = ⁓ =
↓2 plus ↑5 = ⁓ =
↓3 plus ↓1 = ⁓ =
↑7 plus ↓10 plus ↑2 = ⁓ =
↓6 plus ↑1 plus ↑3 = ⁓ =

Do these without the pictures:

↑15 plus ↓20 =   ↑39 plus ↑17 =

↓1 plus ↑500 =   ↑421 plus ↓576 =

↑3000 plus ↓201 plus ↓439 plus ↑752 =

How Do You Do Balloon Problems?

WRITE DIRECTIONS FOR THREE TYPES OF BALLOON PROBLEMS

Be sure to identify **how** to determine if the final effect is ↑ or ↓.

1. If the balloons are all the same type (both ↑ or both ↓.)

2. If the balloons are different types (some ↑ and some ↓), but there are equal amounts of both.

3. If the balloons are different types (some ↑ and some ↓), and there are more of one than the other.

For each of your answers above, circle the part that tells how many balloons in your answer; underline the part that tells how to find out if the effect is ↑ or ↓.

   Teacher Aid 4.2

5. Multiplying Integers

Like many teachers, I use patterning (*Teacher Aid 5.1*) to teach multiplication of integers. My students understand how to do the problems, but struggle to choose words to describe the rules. This activity demonstrates why math books are hard to read.

Multiplying Integers
(Teacher Aid 5.1)

I break the students into cooperative groups of four to complete the Teacher Aid. I then ask them to design a poster with rules for multiplying integers. We want a poster that could be published for a wall decoration for algebra class.

Students turn in these rough drafts. I check them for accuracy and detail.

The next day, they re-write. Then, with felt tips, construction paper and whatever other materials I can put together, I have them make the posters. We display them.

Grading

Students help me grade the posters in class using class discussion. They identify criteria such as clarity or neatness, and we discuss the posters in relation to those student-identified expectations.

Multiplying Integers

We know how to multiply two positive numbers. This worksheet uses patterns to develop multiplication with negative numbers.

FILL IN THE BLANKS AND COMPLETE THE WRITING ACTIVITIES.

$$3 \quad \times \quad 3 \quad = \quad \underline{\hphantom{000}}$$

$$3 \quad \times \quad \underline{\hphantom{000}} \quad = \quad 6$$

$$3 \quad \times \quad 1 \quad = \quad \underline{\hphantom{000}}$$

$$3 \quad \times \quad \underline{\hphantom{000}} \quad = \quad 0$$

$$3 \quad \times \quad \underline{\hphantom{000}} \quad = \quad \underline{\hphantom{000}}$$

$$3 \quad \times \quad \underline{\hphantom{000}} \quad = \quad \underline{\hphantom{000}}$$

$$3 \quad \times \quad (^-3) \quad = \quad \underline{\hphantom{000}}$$

What happens when you multiply a positive and a negative?

$$(^-3) \quad \times \quad 3 \quad = \quad ^-9$$

$$(^-3) \quad \times \quad 2 \quad = \quad \underline{\hphantom{000}}$$

$$(^-3) \quad \times \quad \underline{\hphantom{000}} \quad = \quad \underline{\hphantom{000}}$$

$$(^-3) \quad \times \quad 0 \quad = \quad \underline{\hphantom{000}}$$

$$(^-3) \quad \times \quad \underline{\hphantom{000}} \quad = \quad \underline{\hphantom{000}}$$

$$(^-3) \quad \times \quad \underline{\hphantom{000}} \quad = \quad 6$$

$$(^-3) \quad \times \quad \underline{\hphantom{000}} \quad = \quad \underline{\hphantom{000}}$$

What happens when you multiply two negatives together?

    Teacher Aid 5.1

6. Associative and Commutative Properties

This lesson links the associative and commutative properties with the English meanings of the words. Although the concepts are not difficult, students seem to trip up on the vocabulary. The lesson offers a mental hook to put the vocabulary on.

Opening

I have written the word "associate" on the board. Write down an English word that "associate" make you think of.

I have written the word "commute." Write down what you think it means in English, not math.

Discuss the meanings of the two words. I link "associate" with "group." "Commute" is more difficult for the students. They usually get it when I say that I commute to work. I link "commute" with "travel" or "go back and forth."

Linking the Vocabulary

Today we're working on two algebraic properties that are not surprising in what they do, but we want to acknowledge that they are aspects of our number system. Not all systems work this way.

While we're developing these ideas, be on the lookout not only for when the properties apply, but also for when they do not apply.

It's not going to be difficult for you to believe that these are true properties. The tricky part is to remember their names and to be able to spot them when you see them.

These properties are called the associative and commutative properties.

Teacher Aid
(Teacher Aids 6.1 and 6.2)

These are guided teacher aids, designed to lead the student to a conceptual understanding of associativity and commutativity. The student develops a written explanation of the two concepts as presented on the teacher aids.

Associative Property

Simplify both sides of each equation, showing the work in parentheses first. You will need to write two lines of algebra under each equation.

$(5 + 7) + 3 = 5 + (7 + 3)$	$(2 + 1) + 4 = 2 + (1 + 4)$
$(^-4 + 6) - 3 = ^-4 + (6 - 3)$	$(9 + 7) - 1 = 9 + (7 - 1)$
$(6 \bullet 2) \bullet 3 = 6 \bullet (2 \bullet 3)$	$[(^-3)(2)] \bullet (^-1) = (^-3) \bullet [(2)(^-1)]$
$(^-7 + 8) + 3 = ^-7 + (8 + 3)$	$[(4)(^-3)] \bullet (^-2) = 4 \bullet [(^-3)(^-2)]$

In English, briefly explain what **you** think the associative property is.

Within the associative property, do you think the order of the numbers matters?

Why do you think mathematicians chose the word "associate" to describe this property.

We have used the associative property with addition and multiplication. Explain why you think it will or will not work with division and/or subtraction.

  © 1993 Key Curriculum Press, P.O. Box 2304, Berkeley, CA 94702   *Teacher Aid 6.1*

Commutative Property

Simplify both sides of each equation.

4 + 6 = 6 + 4	7 • 2 = 2 • 7
(500)(100) = (100)(500)	300 + 33 = 33 + 300
23 + 32 = 32 + 23	(⁻1)(⁻5) = (⁻5)(⁻1)
(4 + 5) + 6 = (5 + 4) + 6	(3 • 2)(⁻1) = (2 • 3)(⁻1)

In English briefly explain what **you** think the commutative property is.

Within the commutative property, do you think the order of the numbers matters?

Why do you think mathematicians chose the word "commute" to describe this property?

We have used the commutative property with addition and multiplication. Explain why you think it will or will not work with division and/or subtraction.

7. Solving Linear Equations

I use two activities using writing to learn about linear equations. The beginning activity uses first pictures, then English explanations and finally algebra as ways to represent the solution process. The second activity attempts to help students identify the general steps involved in solving linear equations.

Activity I
(Teacher Aids 7.1a-c)

This is a major assignment in my linear equations unit. It expects the student to use writing and the drawing of pictures to solve equations. **Only after the equations are solved both visually and with an English explanation** will the student represent the solution algebraically.

The algebra goes in the left-hand column. They write that in **after** we have gone over the other answers first.

I emphasize that there is more than one correct way to do the English—students must choose the words themselves. The algebra, however, usually ends up the same.

Writing Activity

I do this additional activity only after students have solved **many** linear equations and are skilled at solving them.

We have a new student in class named Ima Student. Since we have already covered how to solve linear equations, Ima needs to catch up. I want you to help. Write Ima a letter, explaining how to solve linear equations. You may want to include a diagram, number your steps or do something more creative. The important point is that Ima must be able to solve linear equations using your explanation. This is a major assignment. You will write a rough draft that I will read before you do your second draft. **No excuse** spelling words are: multiplication, division, addition, subtraction, equation, and linear.

Equation Worksheet

Think of x as a container holding a certain number.

Example:	$2x - 3 = 7$	
	▽ ▽ ⊖⊖ ⊖ = ⊕⊕⊕ ⊕⊕⊕ ⊕	Original Problem
	▽ ▽ ⊖⊕ ⊕ ⊖⊕ ⊕ = ⊕⊕⊕⊕ ⊕⊕⊕ ⊖⊕ ⊕⊕⊕	Add +3 to BOTH SIDES
	▽ ▽ = ⊕⊕⊕⊕ ⊕⊕⊕ ⊕⊕⊕	Simplify by Removing zero
	▽ = ⊕⊕⊕ ⊕⊕	Take ½ of each side
	$X = 5$	Final Answer

Your turn: fill in the pictures to match each step
$$3x - 1 = 5$$

	Original Problem
	Add +1 to BOTH SIDES
	Simplify by removing zero
	Take ⅓ of each side
	Final Answer

       Teacher Aid 7.1a

Your turn
Fill in the missing parts

$5 = x - 1$	
⊕⊕⊕ ⊕⊕ = ▽ ⊖	
⊕⊕⊕ ⊕⊕⊕ = ▽ ⊖ ⊕	
⊕⊕⊕ ⊕⊕⊕ = ▽	
▽ = ⊕⊕⊕ ⊕⊕⊕	
	Final Answer

$2x - 5 = -1$	
▽▽ ⊖⊖ ⊖⊖ ⊖ = ⊖	
	Add +5 to BOTH SIDES
▽▽ = ⊕⊕ ⊕⊕	
	Take ½ of each side
	Final Answer

Your turn
Do it yourself

$-2 + 4x = 6$	

$x - 4 = -2$	

 Teacher Aid 7.1c

8. Solving Systems of Equations

Before attempting this activity, students need to be comfortable with both the addition and substitution methods of solving systems of equations. These teacher aids (*Teacher Aids 8.1* and *8.2*) allow students to choose which method to use. They also give students an opportunity to clarify how to choose a method. This clarification comes through writing.

Systems of Equations

Solve the following systems of equations, showing your work carefully. Identify the method you use—addition or substitution.

---

(1) $2x + 3y = 13$ $4x + 5y = 23$ x =

y =

Method used:

2) $2y = 8x$ $3x + 2y = 33$ x =

y =

Method used:

(3) $x = y + 5$ $3x - 2y = 11$ x =

y =

Method used:

(4) $x + y = 7$ $x - y = 9$ x =

y =

Method used:

(5) $3 + x = 9y$ $y = {}^-1$ x =

y =

Method used:

(6) $y - 3 = x$ $x + y = 5$ x =

y =

Method used:

       ©1993 Key Curriculum Press, P.O. Box 2304, Berkeley, CA 94702   *Teacher Aid 8.1*

Writing! Writing! Writing! Writing!

SYSTEMS OF EQUATIONS

How do you choose which method to use? Explain your answer carefully. "I pick the easiest," while this may be true, it is not a complete answer. You need to **explain** how you knew one method was easier than the other.

9. Slope of a Line

This lesson relies on student intuition to develop concepts of rise, run and slope. Students use intuition, then develop concepts that reinforce that intuition.

Opening

I'm going to put a word on the overhead. Write down other English words that are linked with it in your mind.

(Show first part of *Teacher Aid 9.1.*)

Discussion

We discuss the student contributions and work into the idea that slope has to do with the slant of a line in mathematics.

You are trusting people. You trust your life on a regular basis to people you do not know. What would happen if you tried to run down these stairs?

(Show center section of *Teacher Aid 9.1.*)

Students will recognize that they would fall.

I ask them to explain what characteristic steps need to have.

They will eventually identify that the steps need to be alike.

It is at this point that I show the final portion of *Teacher Aid 9.1.*

Development/Discussion

(Allow wait time.)

People who build steps pay close attention to what they call the "rise" and the "run" of the steps. What do you suppose that the term "rise" applies to?

What about the other term, "run"?

In order for steps to be built properly, the rise of each step must be the same and the run of each step must be the same. That's why we can run up and down steps; our brains do an automatic geometry problem so that we know **exactly** where that next step will be.

These worksheets (*Teacher Aids 9.2 and 9.3*) develop the concept of slope of a line. Study the lines and the slopes; your job is to figure out how to find the slope of a line.

Students usually do quite well on this activity. They seem to like it, too.

Writing Assignment

Explain how to find the slope of a line. This is not a long writing assignment, but it is tricky. It is not necessary, but may be helpful to use the terms, "rise" and "run." Please write in English—which means full sentences. You may include diagrams or list steps by number if it is helpful, but you need to use language, not just mathematics.

slope

What would happen if you tried to run down these steps?

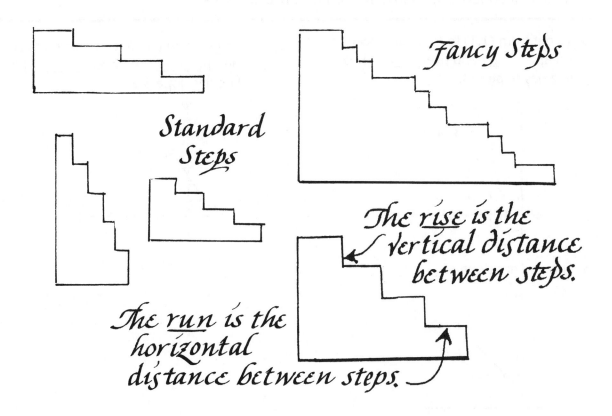

Standard Steps

Fancy Steps

The rise is the vertical distance between steps.

The run is the horizontal distance between steps.

Slope Worksheet

The slope of a line describes its steepness in a given coordinate system. For these exercises, assume the slopes of all the lines are defined by the same coordinate system in which the positive directions are to the right and up from the origin.

1. Since slope has to do with steepness of a line, which of the lines ought to have the largest number for its slope? _____

   (a)  (b)

2. Which of these two lines should have positive slope? _____

   Which should have negative slope? _____

   a) (b)

Mathematicians use numbers to represent slope. In a given coordinate system, steeper lines have larger slope.

Lines like this: / have positive slope.

Lines like this: \ have negative slope.

You can remember this by recognizing that when you travel in the positive direction, you travel uphill if it's a positive slope and downhill if it's a negative slope.

---

3. To get from (1,1) to (4,5), you need to go up _____ and over_____. This line has slope 4/3.

   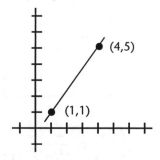

4. To get from (⁻1,1) to (1,4), you need to go up _____ and over _____. This line has slope 3/2.

   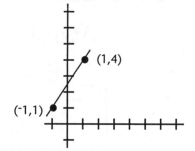

5. This line has slope of _____.

   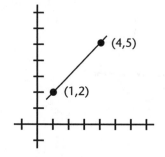

     Teacher Aid 9.2

Find the Slope

1. slope = _____.

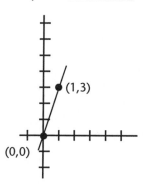

(1,3)

(0,0)

2. slope = _____.

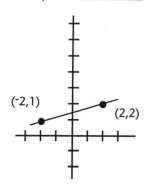

(-2,1)

(2,2)

3. slope = _____.

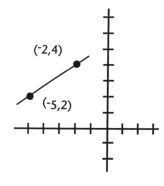

(-2,4)

(-5,2)

4. slope = _____.

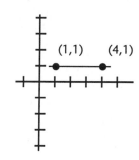

(1,1)   (4,1)

5. slope = _____.

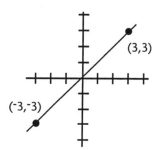

(3,3)

(-3,-3)

6. slope = _____.

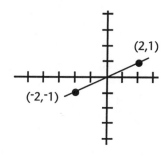

(2,1)

(-2,-1)

7. slope = _____.

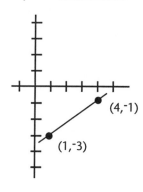

(4,-1)

(1,-3)

8. slope = _____.

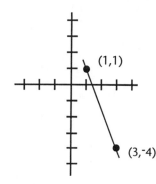

(1,1)

(3,-4)

9. slope = _____.

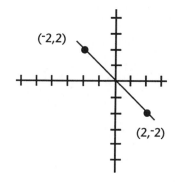

(-2,2)

(2,-2)

10. slope = _____.

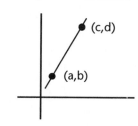

(c,d)

(a,b)

WRITING:

What do the terms "rise" and "run" have to do with slope?

10. Graphing Slope-Intercept Form

Use this graphing activity after teaching about plotting points and slope of lines. Students discover the advantage of slope-intercept form of an equation and identify m and b in equations with form $y = mx + b$.

Opening Activity

Graph the following lines on a piece of graph paper.

A line with slope of 2 that intercepts the y-axis at $^-2$.

A line with slope of $^-1$ that intercepts the y-axis at 4.

A line with slope of 3/4 that intercepts the y-axis at 0.

Go over the students' graphs, checking for prior knowledge. It is important that students understand how to do the opening activity before beginning the major part of the lesson.

Lesson
(Teacher Aid 10.1)

Have students work in groups of two or three for this teacher aid. Emphasize that students must pay close attention to accuracy.

Cruise the room, checking student work as you go.

Pre-Writing/Discussion

Go over student answers to the teacher aid, allowing for discussion of the obvious discovery.

> ### Writing
>
> Think of the equations for the lines as being in the form $y = mx + b$.
> What effect do the m and the b have upon the graph of the line?
> How could you graph the line $y = {}^-2x + 4$ without plotting points?
> Explain your answer in a paragraph. You may want to include some mathematics.

Follow-Up

What are the equations of the lines that you graphed in the opening activity?

Graphing Slope-Intercept Form

Plot at least three points for each of the following equations, and graph the lines carefully. It is important that you graph them carefully so that the rest of the lesson will make sense.

$y = 2x + 3$	$y = 3x + 3$	$y = {}^-4x + 3$
$y = 2x + 1$	$y = 3x + 1$	$y = {}^-4x + 1$
$y = 2x$	$y = 3x$	$y = {}^-4x$
$y = 2x - 1$	$y = 3x - 1$	$y = {}^-4x - 1$
$y = 2x - 3$	$y = 3x - 3$	$y = {}^-4x - 3$

On your graph paper,

Identify the slope of each line.

Identify where each line intercepts the *y*-axis.

11. Function—A Dependent Relationship

It seems as though my students understand the concept of function better and have had an easier time learning dependent and independent variable concepts since I've used these two writing activities. I use the first writing activity as a preliminary to discussions about functions. It seems to give students a better grasp of the concept of function. I am careful to link the first activity to functions by using the "depends upon" language. The second activity helps to solidify the concept for students.

Opening

Use English to complete this sentence:

_____ depends upon _____ .

Examples:

Amount of change depends upon the type of coins.

How green the grass is depends upon the amount of fertilizer.

Weight depends upon calories consumed.

Students write these statements with markers on card stock and post them where everyone can see them. I like posting them because it magnifies the number of examples, and reminds students of the base concept of function.

We develop the concept of function and dependent relationship. After students work with functions for a few days, I return to the "depends upon" idea with the following writing activity.

Applying the Idea

Before we talked about functions, we wrote some "depends upon" statements—statements like "Size of bubble depends upon amount of gum." What did this idea have to do with functions— functions like $y = 2x$? It's not exactly the same, since knowing the amount of gum will not necessarily cause you to be able to predict the size of the bubble. Discuss the similarities of the English "depends upon" statements and the math concept of function.

12. Area is **Not** a Function of Perimeter

The placement of this activity depends upon your curriculum. The lesson attempts to dispel the notion that knowing the perimeter of a region, one can determine the area.

This lesson is on a self-explanatory teacher aid (*Teacher Aid 12.1*). I have students work in partners or even groups of four in completing the pre-writing portion.

Is Area a Function of Perimeter?

Pat needs to find the area of the region shown below.

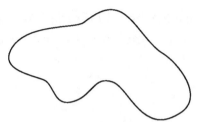

Unfortunately, Pat doesn't quite know how. But Pat has an idea! Pat is cutting a loop of string exactly the same length as the perimeter of the region. Pat intends to re-shape the string as the perimeter of a rectangle. Since finding the area of a rectangle is easy, Pat figures the problem is a cinch.

Pat is absolutely mistaken! Cut a piece of string and make a closed loop. Make a rectangle with the string and determine the area. Make a different rectangle with the same loop and determine its area. Try a third rectangle.

WRITING

Two 1-page essays

1. Explain to Pat why the clever idea for determining area will not necessarily give the true area. Be specific in your answer. **Convince** Pat that the method is inaccurate.

2. Determine a way to come close to the area of the region. There are many correct ways to approach this problem. Discuss it with your partner(s) and explain your method.

 Teacher Aid 12.1

13. Squares and Square Roots

I begin the concepts of square and square root with a manipulative and visual approach. I used this activity quickly with my calculus students. They indicated that it gave them a much better picture. I have students work in groups of two or three. Each group is given 40-50 square tiles.

Opening

Here are some tiles. Some numbers of tiles can be used to make a square.

1 makes a square.

2 can make a rectangle, but not a square.

3 doesn't make a square either.

4 can make a square.

Find the first 10 numbers that can make a square.

Students work on this while I cruise the room to offer any assistance. Some students are quick with this; some students need more time. I ask the quicker ones to determine the 25th square number and justify how they know it's the 25th square number.

Discussion/Pre-writing

What is the English meaning the term "root"? (base of a plant, foundation, . . .)

Now, 3 is the square root of 9; 9 is the square of 3.

What does this tile activity have to do with the terms "square" and "square root"?

Why do you think mathematicians chose the term "square root"?

Why do you think mathematicians chose the term "square"?

Writing

How can I use the idea of which numbers of tiles make a square and the meanings of "square" and "square root" to remember the mathematical meanings of square and square root?

14. Advanced View of Adding Like Terms— Radicals/Fractions/Variables

I don't usually have students discover and write at the same time, but I like the way this particular activity works. It is an attempt to give students a global view of parallels in mathematics, helping students to weave a broader personal scheme. The entire activity is developed in a teacher aid format.

Advanced Like Terms

Section I

$$2/3 + 5/3 = 7/3 \qquad 29/37 + 6/37 = 35/37$$
$$2/5 + 2/5 = 4/5$$

Section II

$$2\sqrt{3} + 5\sqrt{3} = 7\sqrt{3} \qquad 2\sqrt{5} + 2\sqrt{5} = 4\sqrt{5}$$
$$29\sqrt{37} + 6\sqrt{37} = 35\sqrt{37}$$

Section III

$$2x + 5x = 7x \qquad 2x + 2x = 4x$$
$$29x + 6x = 35x$$

Section IV

$$2\pi + 5\pi = 7\pi$$
$$2\pi + 2\pi = 4\pi \qquad 29\pi + 6\pi = 35\pi$$

What is being added in:

Section I   Section II
Section III   Section IV

Fill in the blanks

$$3/7 + 2/7 = \underline{\quad} \qquad 4/13 + 7/13 = \underline{\quad}$$
$$3\sqrt{7} + 2\sqrt{7} = \underline{\quad} \qquad 4\sqrt{13} + 7\sqrt{13} = \underline{\quad}$$
$$3x + 2x = \underline{\quad} \qquad 4x + 7x = \underline{\quad}$$
$$3\pi + 2\pi = \underline{\quad} \qquad 4\pi + 7\pi = \underline{\quad}$$

Writing:
Compare and contrast these four types of problems.

     Teacher Aid 14.1

16. Completing the Square

This activity is different from the others in this book in that it requires little teacher direction. Students prepare to complete the square with five guided-learning worksheets. I like this activity because the students not only "discover" mathematics, they also develop writing that requires the use of algebraic vocabulary. I think it helps students to realize the necessity of vocabulary in algebra.

I usually assign the first worksheet (*Teacher Aid 15.1*) as a homework activity after a test. Students cover the second and third worksheets (*Teacher Aids 15.2* and *15.3*) in class and as homework the next night. The final two worksheets (*Teacher Aids 15.4* and *15.5*) are covered in class.

Completing the Square (1)

COLLECTING INFORMATION:

$(x + 1)^2 =$ $(x - 1)^2 =$

$(x + 2)^2 =$ $(x - 2)^2 =$

$(x + 3)^2 =$ $(x - 3)^2 =$

$(x + 4)^2 =$ $(x - 4)^2 =$

$(x + 5)^2 =$ $(x - 5)^2 =$

$(x + 6)^2 =$ $(x - 6)^2 =$

$(x + 7)^2 =$ $(x - 7)^2 =$

$(x + 8)^2 =$ $(x - 8)^2 =$

$(x + 9)^2 =$ $(x - 9)^2 =$

---

1. How many terms does a binomial have?

2. When a binomial is squared, how many terms are in the result?

3. Do you think that squaring a binomial will always result in the number of items you listed in question 2?

4. **Explain** why you think your answer to question 3 is correct.

     Teacher Aid 15.1

Completing the Square (2)

ORGANIZING INFORMATION

Use the information from worksheet 1 to complete the chart.

BINOMIAL SQUARED	=	RESULTING TRINOMIAL	TRINOMIAL'S 2nd TERM COEFFICIENT	3rd TERM OF TRINOMIAL
$(x + 10)^2$	=	$x^2 + 20x + 100$	20	100
$(x + 1)^2$	=			
$(x + 2)^2$	=			
$(x + 3)^2$	=			
$(x + 4)^2$	=			
$(x + 5)^2$	=			
$(x + 6)^2$	=			
$(x - 1)^2$	=			
$(x - 2)^2$	=			
$(x - 3)^2$	=			
$(x - 4)^2$	=			
$(x - 5)^2$	=			
$(x - 6)^2$	=			

Completing the Square (3)

FINDING PATTERNS

Use worksheet 2 to answer the questions below.

Write your answers in full sentences.
Write your answers in full sentences.
Write your answers in full sentences.

---

1. How are the binomials alike?

2. How are the binomials different?

3. Compare the binomial and the coefficient of the trinomial's second term. What is the pattern?

4. Compare the binomial and the square root of the trinomial's third term. What is the pattern?

5. Using the Information you have deduced above, write a general rule for squaring a binomial such as $(x + b)$. Write your answer in mathematics or full sentences, whichever you prefer.

 Teacher Aid 15.3

Completing the Square (4)

PRACTICING THE PATTERNS

We have concluded the following information from worksheets 1-3:

Squaring a binomial like $(x + b)$ results in a trinomial.

The trinomial has a second term coefficient that is twice the binomial's second term.

The third term of the trinomial is the square of the binomial's second term.

FILL IN THE BLANKS:

$(x + 12)^2 = x^2 + 24x + $ _____

$(x + 11)^2 = x^2 + 22x + $ _____

$(x + 9)^2 = x^2 + 18x + $ _____

$(x + 13)^2 = x^2 + $ _____ $ + 169$

$(x - 4)^2 = x^2 - $ _____ $ + 16$

$(x + 5)^2 = x^2 + 10x + $ _____

$(x + 8)^2 = x^2 + 16x + $ _____

$(x + 7)^2 = x^2 + 14x + $ _____

$(x + 14)^2 = x^2 + $ _____ $ + 196$

$(x - 11)^2 = x^2 - $ _____ $ + 121$

FILL IN THE SIGNS:

$(x + 3)^2 = x^2 \quad\quad 6x \quad\quad 9$

$(x - 9)^2 = x^2 \quad\quad 18x \quad\quad 81$

$(x + 1)^2 = x^2 \quad\quad 2x \quad\quad 1$

$(x + 4)^2 = x^2 \quad\quad 8x \quad\quad 16$

$(x - 5)^2 = x^2 \quad\quad 10x \quad\quad 25$

$(x - 1)^2 = x^2 \quad\quad 2x \quad\quad 1$

SQUARE THE BINOMIALS:

$(x + 6)^2 = $

$(x - 12)^2 = $

$(x + 20)^2 = $

$(x - 10)^2 = $

$(x - 13)^2 = $

$(x + .5)^2 = $

TRY IT THE OTHER WAY:

$x^2 - 16x + 64 = ($ $)$

$x^2 + 18x + 81 = ($ $)$

$x^2 + 60x + 900 = ($ $)$

Completing the Square (5)

PERFECT SQUARE TRINOMIALS

Fill in the blanks, writing a perfect square trinomial. Determine the corresponding binomial.

$x^2 + 8x +$ _____ $= (x +$ _____ $)^2$

$x^2 + 10x +$ _____ $= (x +$ _____ $)^2$

$x^2 - 6x +$ _____ $= (x -$ _____ $)^2$

$x^2 - 14x +$ _____ $= (x -$ _____ $)^2$

$x^2 + 30x +$ _____ $= (x +$ _____ $)^2$

$x^2 + 12x +$ _____ $= ($)

$x^2 + 4x +$ _____ $= ($)

$x^2 - 4x +$ _____ $= ($)

$x^2 - 12x +$ _____ $= ($)

$x^2 - 20x +$ _____ $= ($)

In clear, simple English, explain how to solve the first problem above. Use your own words, not the book's. Make sure your answer is complete.

Teacher Aid 15.5

16. Distortion Project

I include this distortion project when we do graphing. It is a learn-by-doing project that also helps to relieve some classroom boredom. Students who have never shone before often light up for this activity. The project consists of an enlargement and two distortions. Students use writing to explain the process used in developing the reproductions. The project is explained in *Teacher Aids 16.1* and *16.2*. An example of student work is found on pages 100 to 104.

Materials List

Graph paper of various sizes

Tape, stick glue, rubber cement

Clean, white paper

Felt-tip coloring pens

Cartoon pictures

Rulers

Scissors

Distortion Project (1)

Your project will consist of an original picture, three reproductions, and writing. You will be graded on accuracy, clarity of expression, creativity, and neatness.

Trace the three reproductions onto clean, white paper and color them. How to reproduce the picture is explained in the handout.

Each reproduction needs a written explanation of how you made it. Include information on: how you made the grid, how large it is in comparison with the original, special techniques you may have used, frustrations you may have encountered.

DUE DATES

Bring your picture for section I to class this Friday. We will work in class to complete this portion of the project.

The first reproduction will be due the next Friday. The second reproduction will be due the Friday after that. The last reproduction will be due the Friday thereafter.

On each of the three Fridays that a reproduction is due we will do the writing that will accompany it. We will write during class and help each other edit the material.

After all the pictures are finished and graded, you will put the whole project together and turn it in as one body of work. It will be graded as a whole.

 Teacher Aid 16.1

Distortion Project (2)

SECTION I

Find or draw a simple picture. Color books are a great source.

Tape your picture to graph paper.

Use the lines to make a grid over your picture.

SECTION II

Make a grid of larger (or smaller) squares and reproduce your picture changing its size

Trace your new picture onto clean white paper. Color it.

SECTION III

Reproduce your picture as in Section II, but use non-square rectangles to make a distortion.

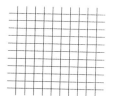

SECTION IV

Reproduce your picture as in Section II, but do something creative that will make your grid non-regular.

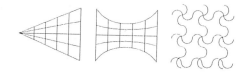

17. Patterning Project

This project idea is adapted from *Designs from Mathematical Patterns* by Bezuska, Kenney and Silvey, Creative Publications, Palo Alto, California, 1978. This excellent book is well worth owning.

This unit uses transformational geometry and modular arithmetic to produce geometric designs and concludes with a writing activity. It's an easy-going end of the year project that students seem to enjoy. This project works well as a two-week culminating activity for algebra. I like the way it ushers algebra students into the summer with an enjoyable forward look toward geometry.

Overview

The student will produce a product consisting of three parts:

1. A modular arithmetic table

2. A design developed from the mod table using transformational geometry

3. A written explanation of the process

Week 1: Transformational Geometry

Day 1: Translations and rotations

Day 2: Reflections

Day 3: Symmetry—rotational and reflection

Day 4: Putting it all together

Day 5: Writing, identification, student examples

Week 2: Modular Arithmetic, Patterns and Projects

Day 6: Modular arithmetic

Day 7: Making a table and choosing colors

Day 8: Developing a basic pattern

Day 9: Designing a project

Day 10: Completing project and writing

Explanation of Project

Students learn transformational geometry moves of translation, rotation and reflection. They use these moves with modular arithmetic to make a design. Each element in the mod set is represented by a color which determines the beginning basic design that the students use. Students are encouraged to be creative, representing the modular pattern visually. Frequent writing activities force the student to analyze and participate.

Materials List

    One copy of each teacher aid per student

    Graph paper

    Felt-tip coloring pens

    Plain white paper for projects and tables

    Colored art paper for backing the projects

    Rubber cement

    Copies of examples

    Playing cards (large size, if possible)

    Pictures of quilt patterns and quilts

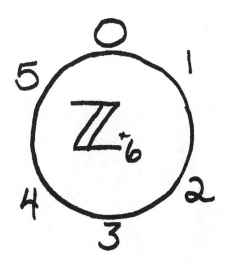

+	0	1	2	3	4	5
0	0	1	2	3	4	5
1	1	2	3	4	5	0
2	2	3	4	5	0	1
3	3	4	5	0	1	2
4	4	5	0	1	2	3
5	5	0	1	2	3	4

In making my pattern I chose a base of six. which means you have six boxes across and six down in each of the four planes. I used the reflection method

+	0	1	2	3
0	0	1	2	3
1	1	2	3	0
2	2	3	0	1
3	3	0	1	2

I used modular arithematic base 4. After coloring one 4x4 square I reflected it four times then reflected that four times.

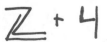

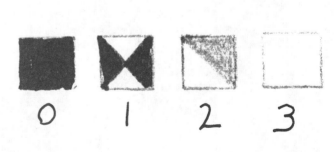

0   1   2   3

rvis Williams
r. 6

Patterning Project, Day 1

Translations and Rotations

Objective

The student will be able to identify and develop geometric translations and rotations.

Opening Activity

(Teacher Aid 17.1a)

Show the translation and rotation patterns on the overhead. Give students about three minutes to complete the writing activity.

Discussion

To *translate* is to move something from one language into another. The goal of the translator is to move the piece from one language to another without changing it. A geometric translation moves the original form, but does not change it in any way—not even by turning it. A pattern like pattern A is called a translation.

To *rotate* is to turn about an axis or a point; this is exactly what we did in pattern B. Patterns made in this way are called rotations.

Lesson

Students read the results of the opening activity writing to the class. You may want to have about ten students read and then ask for additional volunteers if students think there is yet another way to describe the patterns. Emphasize key phrases by writing them on the board. Discuss differences. Develop a class definition for each of the two patterns.

Practice

(Teacher Aid 17.1b)

On this page are some translations and some rotations. Identify each pattern as either a translation or a rotation.

Assignment

Make a capital *I* on your paper. Use this letter and make two patterns with it. One pattern will be a translation and one will be a rotation. Write an explanation of what you did to make each of the patterns.

Patterns

Beginning with the above basic shape, we can make two basic patterns ———.

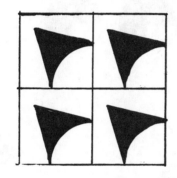

WRITING:

Use your own words to describe how each pattern was made from the original basic design.

Translations and Rotations

Which designs are translations?

Which designs are rotations?

Which designs are both translations and rotations?

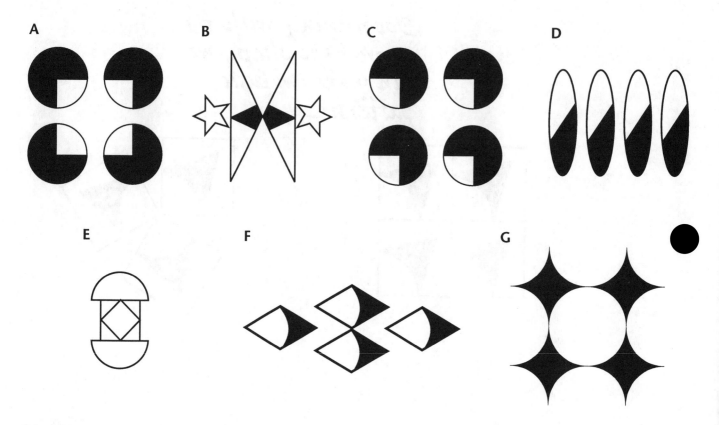

A B C D

E F G

WRITING

Write about another way to make a pattern other than by translation or rotation.

 Teacher Aid 17.1b

Patterning Project, Day 2

Reflections

Objective

The student will be able to identify and develop patterns using reflections.

Opening Activity
(Teacher Aid 17.2a)

Show the reflection patterns on the overhead. Have students take about three minutes to complete the writing activity.

Lesson

Students read their opening activity writing. Emphasize key phrases by writing them on the board. Develop a class definition for the pattern.

Discuss English meaning of the word "reflect," linking it to the geometric concept of reflection. If you could place a mirror on the picture and get the other half of the picture in the mirror, it's a reflection.

Practice
(Teacher Aid 17.2b)

On this page are ways to practice the concepts of translation, rotation, and reflection.

Translations, Rotations, Reflections

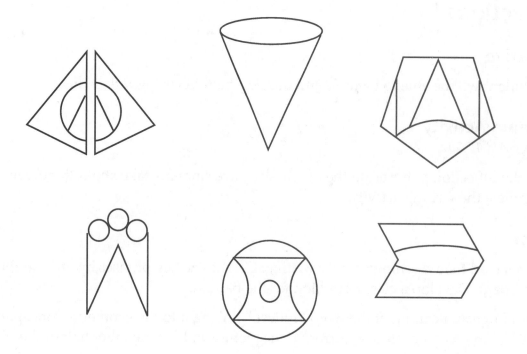

The six patterns above could all be made using the same method. Describe the method.

    Teacher Aid 17.2a

Transformations

1. Make a non-symmetric design.

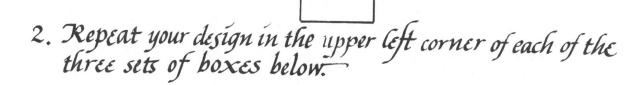

2. Repeat your design in the upper left corner of each of the three sets of boxes below.

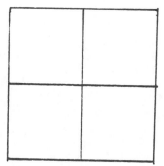 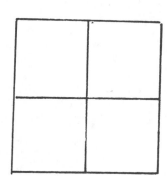

3. Translate   4. Rotate

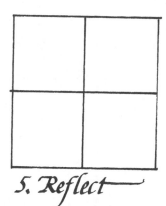

5. Reflect

Patterning Project, Day 3

Symmetry—Rotational and Reflection

Objectives

The student will be able to identify both reflection symmetry and rotational symmetry. The student will be able to determine lines of symmetry.

Opening Activity

Give students poker chips or colored blocks and ask them to make a pattern.

Lesson

Using overhead chips, duplicate some of the patterns.

Have students identify which patterns they think are symmetric and which are not.

Define line of symmetry. Ask students to give examples. Responses are virtually unlimited and will include such as the following: ink blots, architecture, tables, chairs. Identify examples in the room.

Show quilt patterns. (I use pictures from a quilt calendar.) Have students identify what constitutes an individual pattern. (Answers will vary.) Have students determine lines of symmetry.

Using *Teacher Aids 17.1a* and *17.2b* (used previously on days one and two), draw lines of symmetry on an overhead transparency.

Practice

List which letters of our alphabet have a line of symmetry. These letters will have what we call "reflection symmetry."

Exploration

See if you can find some letters with what we call "rotational symmetry." We have not discussed rotational symmetry; use your intuition.

Discuss practice answers. Discuss rotational symmetry.

Assignment
(Teacher Aid 17.3)

Symmetry worksheet.

Lines of Symmetry

Draw all lines of symmetry.

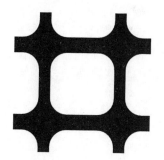

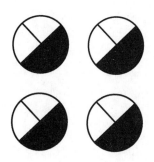

Patterning Project, Day 4

Putting It All Together

Objectives

The student will be able to identify patterns involving translations, rotations, reflections.

Students will be able to determine rotational and reflection symmetry.

Teaching Note

This day is dedicated to putting together what the students have learned during days one, two and three. Students see many types of patterns at once and are able to clarify weak personal definitions of the specific geometric concepts.

Opening Activity

Have students list as many of the geometric concepts as they can remember. After they write their own lists, make a group list using input from the class. It should include the terms, translation, rotation, reflection, rotational symmetry, and reflection symmetry.

Lesson

Discuss the meaning of each of the terms listed in the opening activity.

Using *Teacher Aid 17.4*, have students identify as many of the patterns as possible. Elicit student identification of some patterns as being more than one type.

Using *Teacher Aid 17.4*, have students work in groups of three, developing written definition of the terms: translation, rotation, reflection, rotational symmetry, and reflection symmetry.

Go over student answers by having each group report and lead class discussion of a section of the activity.

Assignment

Prepare for a test tomorrow by:

- making an outline of what we've covered this week and
- writing 10 questions about what we have covered this week.

More Patterns!

A

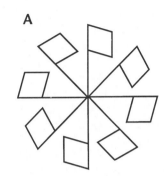

B

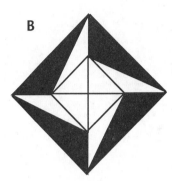

Which patterns could be made using:

translation?

rotations?

reflection?

C

Which patterns have rotational symmetry?

D

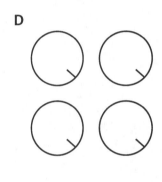

Which have reflection symmetry?

E

F

G

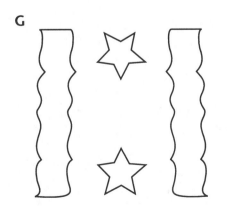

H

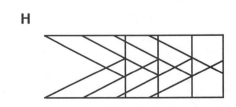

Patterning Project, Day 5

Writing, Identification, Student Examples

Objective

To be able to identify and develop examples of translations, rotations, reflections, rotational symmetry, and reflection symmetry.

Opening Activity and Test Preparation

Post examples of student work from the Lines of Symmetry assignment on day three. Discuss why some students may have identified more lines of symmetry. Summarize material from days three and four. Allow time for student questions.

Give Test
(Teacher Aid 17.5)

At the front of the room, post the following playing cards:

> 10-hearts, 10-diamonds, 3-spades, 7-clubs, 5-hearts, 5-clubs, Q-diamonds, J-clubs, K-spades, K-hearts

Transformational Geometry Test

1. Choose a letter of the alphabet that has neither reflection symmetry nor rotational symmetry.

2. Using the letter you chose in problem 1 above, make a pattern that has reflection symmetry.

3. Using the letter you chose in problem 1 above, make a pattern that has rotational symmetry.

4. Some playing cards are posted on the board. Find an example of each of the following concepts:

    translation

    rotation

    reflection

    rotational symmetry

    reflection symmetry

5. For each of the concepts listed in problem 4 above, convince me that your example fits the concept by justifying your answer carefully. Use the back of this paper.

 Teacher Aid 17.5

Patterning Project, Day 6

Modular Arithmetic

Objective

To be able to perform addition and subtraction using modular arithmetic.

Opening Activity

I'm going to show you a kind of arithmetic you use every day. Your job is to find out what it is.

(Show "clock" arithmetic, using random choices.)

$4 + 7 = 11$ $11 + 6 = 5$ $9 + 5 = 2$ $8 + 10 = 6$

Students can suggest problems, with the teacher supplying the answer until the method is discovered. Suggestions such as 9 + 15 go in a column under "I can't do it." Continue doing student suggested problems until they identify that you're computing time.

Lesson

See example and chart on *Teacher Aid 17.6*.

Model modular addition, base 4. Important Points: Base 4 implies four elements (0, 1, 2, 3). There is no 4 in the set because we begin with 0.

Drawing a picture of a clock with the numbers 0, 1, 2, and 3 is helpful to students. They will find this visual particularly helpful later when the calculations are more complex.

Make a table for addition, mod 4.

Have students find patterns—as many as they can. List them.

Repeat above process for subtraction, mod 4.

Note: Addition has a diagonal pattern with a positive slope; subtraction has a diagonal pattern with a negative slope.

Assignment

1. Make a table for addition, mod 5 and a table for subtraction, mod 5.

2. Write a letter to your best friend explaining how to do modular addition. Be careful to include enough detail that your friend can do modular addition without having to ask you any questions.

Modular Addition · Base 4

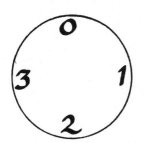

+	0	1	2	3
0				
1				
2				
3				

2 + 3 means:

Start at 2 and move 3 steps in a clockwise direction.

Thus, 2 + 3 = 1

How would you define subtraction?

Patterning Project, Day 7

Making a Table and Choosing Colors

Objectives

The student will be able to perform modular multiplication.

The student will determine a base and an operation for a project.

Opening Activity

Divide students into groups of three. Ask them to determine a way to perform modular multiplication and prepare a table for multiplication, mod 4 *(Teacher Aid 17.7a)*.

Lesson

Discuss student answers.

Determine a table for mod 4 multiplication.

Do mini-project on *Teacher Aid 17.7b*.

Overview of Project

(See examples.)

This project uses geometric patterning and modular arithmetic to make colored patterns. The student chooses a particular modular arithmetic operation and base and selects colors and/or patterns to represent each number in the set. The operation table is represented with the color and/or pattern assigned to each number. This color-pattern chart will be the student's basic pattern. It is the student's responsibility to reproduce the basic pattern, making a design that uses the basic pattern at least four times. This final design will be developed using transformational geometry. See examples. The final project will include a written explanation of how the entire project was done.

How to Begin

Each student must select a modular base and an operation. I caution students that selecting a small base is fine, but a small base needs a more complicated final project than a larger base. I do not, however, try to discourage the use of small bases. I find that these projects can be wonderful.

After selecting the base and operation, each student needs to make a neat operation table for the project.

Under the operation table the student makes a color pattern key. Each number in the modular set will be represented by a color and/or pattern.

Assignment

Using the base of your choice, complete an operation table and key. Reproduce the arithmetic table using the color patterns chosen. The table, key, and reproduced table should all be placed neatly on the same page.

Modular Multiplication

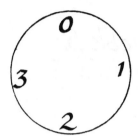

·	0	1	2	3
0				
1				
2				
3				

Develop a method for modular multiplication. Explain it.

Modular Design

	0	1	2
0			
1			
2			

Fill in the table for subtraction, mod 3.

Make a different design in each of the three blocks.

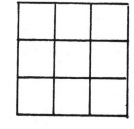

0   1   2

Using your designs for 0, 1, 2 and the subtraction table above, fill in the first nine squares with designs.

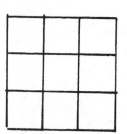

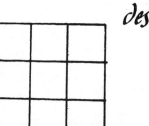

Rotate the patterns to complete the design.

     Teacher Aid 17.7b

Patterning Project, Day 8

Developing a Basic Pattern

Objective

The student will design a project.

Opening Activity

Show examples of student projects.

Lesson

Go over any questions about yesterday's activity.

The student needs to develop a grid on which to place the reproductions of the pattern. Model a sample project (see *Teacher Aid 17.8a*).

Assignment

Develop a grid that is appropriate for your modular arithmetic table using that grid. Remember to make it large enough to use transformational geometry and repeat your pattern at least four times. The projects will be evaluated based on complexity and quality of presentation. (Neatness counts!)

Patterning Project

1. Pick a base and an operation. Make the table. (Using graph paper will keep it neat!)

2. Make a grid that fits your table. You may use graph paper to make a regular grid or do something creative as in the distortion project. Your grid needs to be repeated at least four times or be four times as big as your table.

3. Choose a color and/or pattern for each of the numbers in your set.

4. Fill in the grid using your mod table as a guide. Repeat your pattern at least four times, using at least one geometric transformation.

5. Your project will include a written explanation. It must include details as to the base and operation, tell how you made your grid, and relate the geometric transformations used in repeating the pattern.

 Teacher Aid 17.8a

Patterning Project, Day 9

Designing a Project

Objective

The student will represent a modular arithmetic table using color, pattern, and transformational geometry.

Lesson

Have materials available for completion of the project (coloring pens, white paper, graph paper, colored art paper for backing projects, rubber cement.) Students work on projects. This usually is quite an informal day, students and teacher working and chatting.

Assignment

Work on your project and be ready to mount it and turn it in tomorrow.

Patterning Project, Day 10

Completing the Project, Writing

Objectives

The student will complete the transformational geometry and modular arithmetic project.

The student will also describe the process.

Pre-Writing Activity

You need a partner. Explain your project to your partner. Be sure to cover the following:

1. What base and operation did you use?
2. What color and/or pattern represents each number?
3. How did you use transformational geometry to repeat your base pattern?

Writing

Explain your project in writing, paying careful attention to detail. Be sure to cover the three points we discussed above.

Lesson

Working in groups of three, each student reads what s/he wrote while the other group members look at the project. Students help each other to come up with a clear explanation of each project.

Students copy the writing on clean, white paper, arranging the writing, operation table, color-pattern key, and copy of the base pattern on the paper. Students cut out final designs, mounting them on the colored art paper with rubber cement.

The paper with the writing is stapled on as a second sheet.

18. Additional Writing Topics and Activities

I have found that in order for algebra students to write successfully, the teacher must be aware of the pre-writing atmosphere. It's important to develop topics so that students have something to write about. Sliding into a writing activity works well. Hitting them cold with a topic does not seem to work. You need to warm up the motor before beginning. The following topics and activities must be preceded by a warm up—a pre-writing experience to help them develop ideas about the topic.

1. Math Journals

Many teachers have their students keep a daily journal in math class. Students usually write during the last 10 minutes of the period. The topic is often open for students to choose.

2. Timed Writing

Students write continuously on a topic for five minutes. If they cannot think of what to write, they are to copy the topic over and over until they think of something. The idea is to let thinking flow without restricting it.

3. Question Writes

Students write for five minutes and end with a question. A question a day is an interesting activity. Students learn by choosing questions.

4. Advice

Students write to next year's algebra students or imagined new students to this year's class.

5. Analysis

Students analyze errors from a recent test. Students can also analyze a teacher-offered solution to a problem, finding errors and explaining to another student.

6. Directions

Students write directions for a process. I encourage them to format their directions creatively. Writing in paragraph form is not always a good idea.

7. Essay Questions

Students write essay test questions. Students can also write questions for teacher-led or student-led discussion.

8. Commercials

Students write commercials, selling a math technique.

9. Tell Me What You Know

I met a teacher (name unknown) who gives two tests on every chapter. Both tests count equally for grading. The first test is a traditional math test. The second test is, "Tell me everything you know about Chapter X." Students score one point for every true fact and lose two points for every false fact listed. The tests are graded on a curve.

18. Additional Writing Topics and Activities

19. Examples of Student Writing

Below are excerpts of actual, unedited student writing based on the activities in this book.

Chapter 1. I Hate Math

Things a Math Teacher Should Do

a. Try to be relaxed with a class. If your comfortable the class will be comfortable too.

b. Be tough but not too tough and don't give homework on Fridays.

c. Go slow when doing notes. Give less homework—make it harder. Help kids individually, if they need it.

d. They need to know why! If people are told to do something tell them why.

Things a Math Teacher Should Not Do

a. Never let them see you sweat.

b. Don't have the same routine. Think of ways your class can have fun learning about what your teaching.

c. Don't always yell at student, especially in front of class. If you always jump on them, they will dearly hate you, deary, and do things to stay out of class.

d. Don't always be a shot and have favorites.

Something That Happened in Math Class

a. In fifth grade math teacher asked if I could walk and chew gum at the same time. Naturally, I thought it was his/her way of asking me if I was chewing gum, which I wasn't. So, I said, "No." The whole class heard it, and I was humiliated.

b. "You should act like Wendy, In fact this whole class should be like her. Quiet!"

c. The teacher said I did it a smart way and it made me feel real good.

d. When I was in fourth grade, my mom had to go on a business trip. It was her first such trip, and I was depressed. In my math class, instead of doing my work (I never did anyway) I sat and drew melted flowers and droopy stars. I remember beginning to cry. My teacher took me outside and talked to me,. She told me how her mother always came back. It seems dumb now, but that sweet lady helped a little girl in distress. I still remember.

Chapter 2. What is Multiplication (In 25 words or less)

a. Multiplication is a quick way of adding numbers that are the same. Example: instead of adding 4 + 4 + 4 + 4, you would simply multiply the 4 x 4.

b. Multiplication is a greater, easier and faster way to add a great deal of the same numbers together faster.

c. Multiplication is a faster and easier way of adding the same number many times.

d. Multiplication is the act of adding items of the same value in a quicker, more efficient way.

Chapter 4. Adding Integers

How to do Balloon Problems

a. If they both go the same way, you add the numbers and the balloons go the same way.

b. If they both go different ways, you subtract and they go the way most of them went before.

Chapter 8. Systems of Equations

How to Choose When to Use the Substitution Method and When to Use the Addition Method

a. When one of the equations is $y =$ something or $x =$ something, you can substitute the x or y in the other equation. You use the addition method when things are lined up so you can cancel them out.

b. You use substitution when at least one equation is $x =$ or $y =$. But if both equations are $x + y$? Then you use addition because you could cancel the x or the y out. If you use substitution in this type of equation, you'd have to rearrange it.

Chapter 9. Slope of a Line

The word slope, can also be defined as a slant, grade, inclination, bevel, diagonal, rise, fall and pitch. All of these words mean about the same thing. The slope of a line can be determined in the following way:

First of all, one must find the rise and run of the line. The rise is the number of squares it takes to reach a certain point on the line, either up or down. The run is the number of squares it takes to reach a certain point of the line across. Then, the rise is placed over the run. The outcome is the slope.

Chapter 12. Radicals/Fractions/Variables

Compare and Contrast These Three Types of Problems

Adding radicals or fractions is just like adding variables. If one treats the radical as a variable and adds the coefficients. As long as the variable, radical or denominator is the same, the numbers can be added as usual.

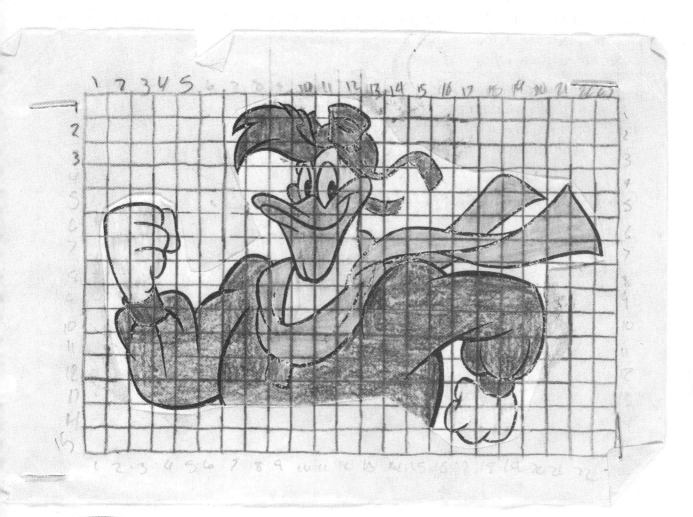

Damon Scarborough

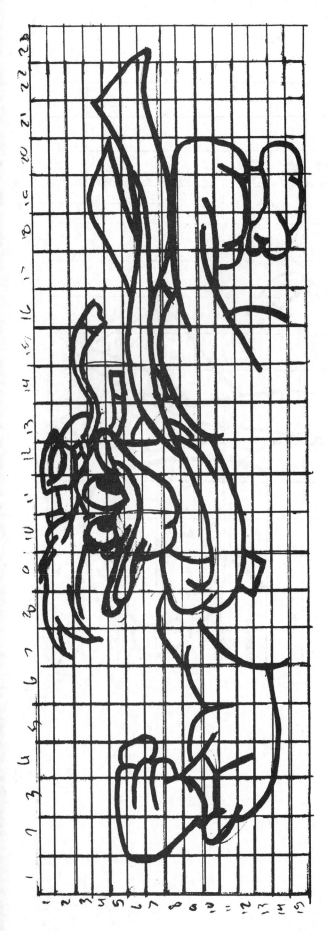

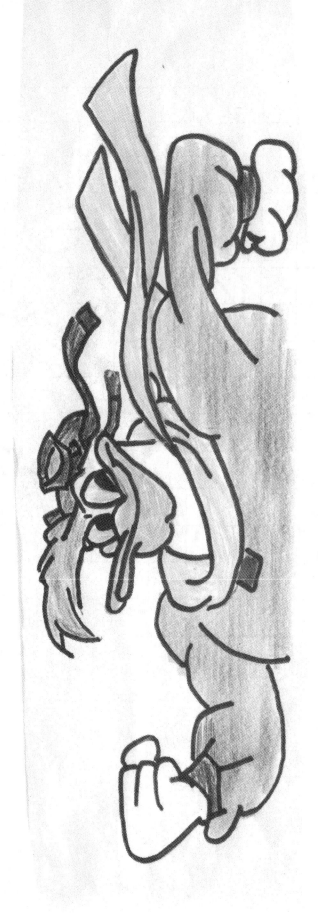

Distortion Project Example

104